Matthew's Sunshine Bakery
Multiplication Arrays

Kathleen L. Stone

Enjoy these other books by Kathleen L. Stone

Penguin Place Value
A Math Adventure

Number Line Fun
Solving Number Mysteries

Riley the Robot
An Input/Output Machine

Mason the Magician
Hundreds Chart Addition

Katelyn's Fair Share Picnic
More Math Fun

Money Tree Mysteries
Adventures with Quarters

Alien Even and Alien Odd
A Math Space Adventure

Kenley's Line Plot Graph
Another Math Adventure

From My Quilted Heart to Yours
*Heart Warming Quilts and Heart Healthy Recipes for
Your Loved Ones*

ISBN–13: 978-1511694490
ISBN-10: 1511694491

Dedication

To my great-niece, Samantha, and great-nephews, DJ, Maurice, Gavin, Jake, Daniel, and Matthew! You bring sunshine to all our lives and we love you all very much.

Matthew the chef
Loves to bake
Cookies, candies,
Pies and cakes.

He meticulously places
Each treat on a tray
Organizing each one
Into a tasty array.

Take a look at these cookies
Matthew baked for afternoon tea.
How would you describe this array?
That's right, *three* rows of *three*!

Now we could say 3 + 3 + 3
But we can save ourselves some time.
Looking at this array let's say
Three times *three* is *nine*.

It really is much easier
When we multiply.
Here's another tray of treats.
Are you ready to give it a try?

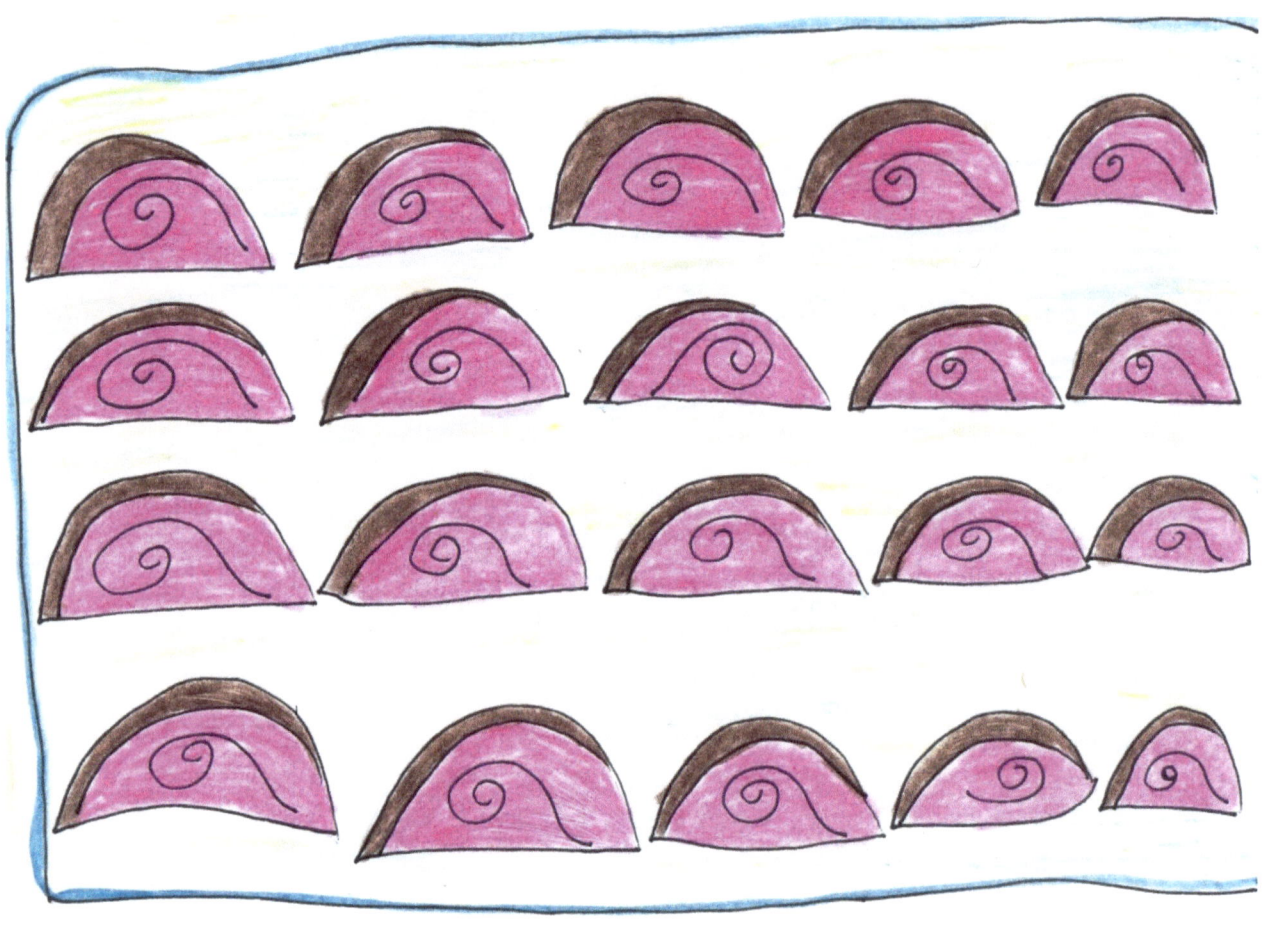

Choco~Bites
15¢ each

There are *four* rows of *five* candies
Arranged on Matthew's tray.
So what's the multiplication equation
You would say for this array?

Did you say *four* times *five* equals *twenty*?
If you did, you were right!
You are a great mathematician.
A quick thinker and very bright!

What multiplication equation
Describes these scrumptious treats?
Five times *three* is *fifteen*.
Mmm, they're such yummy looking
sweets!

Today Matthew is baking
Strawberry tarts for his cousin, DJ.
There are *four* rows of *two* tarts
Arranged neatly on his tray.

4 rows of 2

$2 + 2 + 2 + 2 = 8$

$4 \times 2 = 8$

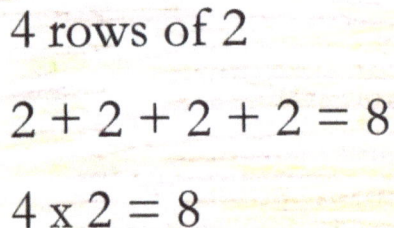

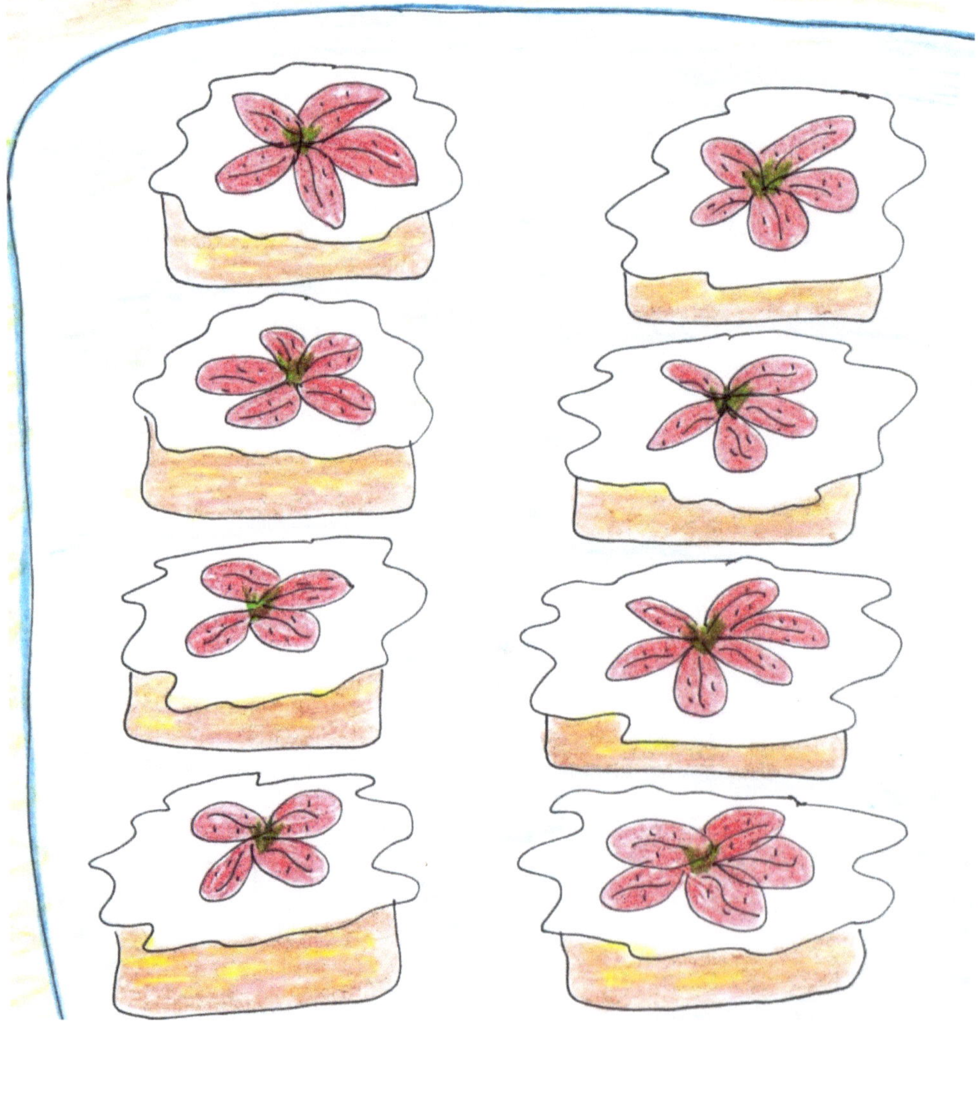

What would be the equation?
Did you say *four* times *two* is *eight*?
That's the correct equation … good
for you!
You are really doing great.

Petite Pistachio
Pies
80¢ each

Matthew's almost done baking
But he'll give you a couple of tries
To figure out the equation
For this tray of pistachio pies.

4 rows of 4

$$4 + 4 + 4 + 4 = 16$$

$$4 \times 4 = 16$$

Four rows of *four* pies each.
Hmmm, *four* times *four* is *sixteen*.
Sixteen delicious pistachio pies
All covered with sweet whipped
cream.

Matthew's yummy day of baking
Has finally come to an end.
But maybe you can practice
Even more multiplication with a
friend.

Arrays

An array is an arrangement of objects, pictures, or numbers in columns and rows. Arrays are useful representations of **repeated addition** and **multiplication** concepts.

Children should be given many opportunities to work with concrete examples (hands on manipulatives) before moving on to more abstract concepts. Encourage children to think of "real life" arrays they have seen … eggs in an egg carton, the "bumps" on Legos, paints in a watercolor tray, crayons in a crayon box, etc.

Enrichment Activities
Hip, Hip Array

Materials needed:

1" graph paper
12 cubes or similar objects
White boards and markers (or paper and pencil)

Using the graph paper as a work mat, children can practice making various arrays using their twelve cubes, and writing the multiplication equations for each array. For example,

- ♥ *three* rows of *four* (3 x 4)
- ♥ *four* rows of *three* (4 x 3)
- ♥ *two* rows of six   (2 x 6)   etc., etc., etc.

You might even choose to give children a square waffle and twelve raisins to make this an edible array activity.

Guess My Array

Assign a homework activity for your students to make a multiplication array at home using their favorite toys, snacks, or any other objects (with the multiplication equation written on the <u>back</u> of their array). Put the arrays on display and have the children go around the classroom, writing multiplication equations for their classmates' arrays!

How to Use This Book

I use *Post It* notes to cover the equations on the illustrations when reading this book to my students. That way they have the opportunity to solve each array first. You may also practice other skills using the illustrations …
What coins would you use to purchase one pistachio pie? How much would three Rainbow Delights cost? What fraction of cupcakes has pink frosting? What would the array be for the strawberry tarts if I turned the picture sideways?

ABOUT THE AUTHOR

Kathleen Stone is a National Board Certified educator and is currently teaching second grade. She loves spending time with her family. She and her husband Gary live in the Olympia area. When not teaching, Kathleen can often be found quilting or sitting by the lake reading!

Math is all around us
No matter where you turn
Open your mind to the wonders of math
And all that you can learn

www.ingramcontent.com/pod-product-compliance
Lightning Source LLC
Chambersburg PA
CBHW050411180526
45159CB00005B/2226